REGIONAL ACTION ON CLIMATE CHANGE

A VISION FOR THE CENTRAL ASIA REGIONAL ECONOMIC COOPERATION PROGRAM

APRIL 2024

Endorsed at the 22nd CAREC Ministerial Conference
30 November 2023
Tbilisi, Georgia

CAREC
Central Asia Regional Economic Cooperation Program

ADB

Note:
In this publication, "$" refers to United States dollars.

On the cover: Nurek hydropower plant in Tajikistan (photo by Nozim Kalandarov); the Bridge of Peace in Tbilisi, Georgia, a bow-shaped pedestrian bridge made from steel and glass that spans the Kura River, linking Rike Park with the old town (photo by Khatia Jijeishvili/ADB); wind turbines along the highway in Georgia (photo by Eric Sales/ADB); daily operations at the 15-megawatt (MW) Sermsang Khushig Khundii Solar plant in Khushig valley, Tuv Aimag in Mongolia (photo by Ariel Javellana/ADB); and herders in Mongolia cultivate fodder or animal feed that is more resilient to extreme weather changes, and using plants that adapt to droughts (photo by Eric Sales/ADB). Cover design by Judy Yñiguez.

Contents

Executive Summary   v

Abbreviations   vii

1   Introduction   1

2   Vision, Goals, and Principles for the Central Asia Regional Economic   3
    Cooperation Climate Action

3   Implications of Climate Change for Regional Cooperation   5

4   Central Asia Regional Economic Cooperation: A Regional Platform   7
    for Climate Action

5   Central Asia Regional Economic Cooperation's Focus on Climate Change   9
    in Strategies and Programs

6   Financing Climate Action   15

7   Guiding and Monitoring Results for Implementing the Climate Change Vision   17

8   Summary of Recommendations for Ministers   19

Appendixes   21
    1   Development Partners' Climate-Related Support to the Central Asia Regional   21
        Economic Cooperation Program
    2   Draft Terms of Reference for the Central Asia Regional Economic Cooperation   24
        Program Climate Change Steering Committee

Glossary   25

References   27

Pakistan. The ADB-supported project in Islamia University Bahawalpur Solar Park enhances Pakistan's energy security by helping install clean energy sources and improves people's access to electricity in Punjab, Pakistan (photo by Rahim Mirza/ADB).

Executive Summary

The effects of climate change are already evident. Member countries of the Central Asia Regional Economic Cooperation (CAREC) Program face higher than average increases and greater variability in temperature, and greater variability and extremes in other weather conditions such as precipitation. These phenomena have resulted in increasing water scarcity, flooding, drought, desertification, migration, and conflict; and declining agricultural productivity, food security, and health.

The years 2022–2023 witnessed particularly dramatic and deadly examples of climate change in the region, including devastating floods in Pakistan, severe drought in Afghanistan, floods in the People's Republic of China, excessive heat beyond historical levels, and transboundary water issues in Central Asia. These trends and climate-linked events are sharp reminders of the long-term likelihood of even more serious impacts if steps are not taken to control carbon emissions and increase countries' resilience. Climate change in the CAREC region requires urgent and collective action.

To limit the impact of climate change and improve resilience, CAREC countries will have to pursue energy efficiency, sustainable agriculture, industrial modernization, and renewable energy, including applying heating and cooling technologies. They will have to invest in their economies through mitigation and adaptation, introducing new mechanisms and financial instruments. The region can do better in the use of water, implementation of efficient transport systems, and creation of sustainable cities. It can build more effective health and education systems, strengthen social security, increase fiscal sustainability, and enhance early warning systems for disaster events. Cooperation through dialogue and collective action among CAREC countries will be essential to address regional linkages and spillovers.

The CAREC vision outlined in this publication includes principles for climate action across sectors. It identifies priority areas for investment, explicitly establishes climate change as a crosscutting priority under the CAREC 2030 Strategy, and proposes steps and institutional arrangements to achieve more sustainable and climate-resilient growth.

Georgia. Wind turbines along the highway emit little to no greenhouse gases or pollutants into the air (photo by Eric Sales/ADB).

Abbreviations

ADB	Asian Development Bank
AIIB	Asian Infrastructure Investment Bank
CAREC	Central Asia Regional Economic Cooperation
DMC	developing member country
EBRD	European Bank for Reconstruction and Development
EU	European Union
ETM	Energy Transition Mechanism
ETS	emissions trading system
GHG	greenhouse gas
IsDB	Islamic Development bank
JETP	Just Energy Transition Partnership
LCP	low-carbon and climate-resilient pathways
MDB	multilateral development bank
NDCs	Nationally Determined Contributions
PRC	People's Republic of China
UN	United Nations
UNDP	United Nations Development Programme

Mongolia. A thin line of smog hovers above the *ger* district in Ulaanbaatar. During winter, when temperatures drop below −30° Celsius, the air quality also drops to hazardous levels because of over-dependence on burning cheap raw coal for heating and cooking (photo by Ariel Javellana/ADB).

1 Introduction

The Central Asia Regional Economic Cooperation (CAREC) Program comprises 11 countries that cooperate for mutual benefit and greater regional prosperity: Afghanistan, Azerbaijan, the People's Republic of China (PRC), Georgia, Kazakhstan, Kyrgyz Republic, Mongolia, Pakistan, Tajikistan, Turkmenistan, and Uzbekistan.[1] All CAREC member countries are severely affected by climate change and will be even more so in the future. These countries differ in terms of land area, population size, geographic characteristics, natural resource endowment, per capita income, human capital development, and institutional capacity. Therefore, they also differ in terms of both their contribution and vulnerability to climate change.[2] Accordingly, many of the solutions to the climate impact in the region will have to be tailored to the conditions and needs of each country and subregion. While every country has its peculiarity of climate change impact, there are important commonalities, regional linkages, and spillovers that make regional approaches and cooperation necessary and appropriate. These will include regional dialogue toward collective actions, such as agenda setting, investments for projects with regional significance, policy coordination, capacity building and research, knowledge exchange, and technology transfer.

The CAREC Program offers a unique regional platform for driving coordinated climate action. Building on over 20 years of successful support to its developing member countries (DMCs)[3] with the assistance of a growing number of multilateral and bilateral development partners, the CAREC Program can be an effective mechanism to address climate change challenges. It can provide technical assistance and investments for regional solutions; establish regional frameworks for enhanced project design and financing; and support regional policy coordination, capacity building, and knowledge solutions.[4]

---

[1] The Asian Development Bank (ADB) placed its regular assistance to Afghanistan on hold effective 15 August 2021.
[2] See ADB. 2023. *CAREC 2030: Supporting Regional Actions to Address Climate Change–A Scoping Study*. https://www.carecprogram.org/uploads/Supporting-Reg.-Actions-to-Address-Climate-Change_WEB-ENG.pdf. Global climate change research stresses that climate change impacts affect different regions, countries and subnational districts differently. NBER. 2022. *Climate Change Around the World*. https://www.nber.org/papers/w30338.
[3] See ADB. 2023. *Evaluation of ADB Support for the Central Asia Regional Economic Cooperation Program, 2011–2022*. https://www.adb.org/documents/evaluation-adb-support-central-asia-regional-economic-cooperation-program-2011-2022.
[4] ADB as the CAREC Secretariat, is fully committed to supporting its DMCs in tackling climate challenges in the region. See Appendix 1 for Development Partners.

Tajikistan. Nurek hydropower plant
(photo by Nozim Kalandarov/ADB).

2 Vision, Goals, and Principles for the Central Asia Regional Economic Cooperation Climate Action

The design and implementation of CAREC climate actions need to be underpinned by a vision, goals, and guiding principles. The proposed vision statement for CAREC climate action is as follows:

CAREC Climate Change Vision Statement:
"A Region of Sustainable Development, Shared Prosperity, and Climate-Resilience"

CAREC will pursue three broad goals under the Climate Change Vision:

- **Mitigate climate change**: CAREC shall explore possibilities to support its member countries to achieve carbon neutrality during the middle of the 21st century by targeting activities at national and regional levels in various areas (energy, transport, agriculture, urban development, etc.) to reduce greenhouse gas (GHG) emissions and increase GHG sequestration, consistent with the United Nations Framework Convention on Climate Change and its Paris Agreement.

- **Adapt to climate change**: CAREC shall explore possibilities to assist its member countries to adapt to actual or expected climate change and its effects and to protect their economies and people by promoting and implementing climate adaptation, disaster risk reduction, and financing measures; protecting natural resources (agriculture, water, biodiversity, and human health); and supporting complementary actions in other sectors (education, social protection, financial services, etc.).

- **Cooperate across borders**: CAREC shall explore possibilities to achieve mutually beneficial climate and development outcomes by cooperating in areas with strong cross-border linkages and spillovers and by sharing technologies, knowledge, and experience in all areas of climate action.

Consistent with the principles of the CAREC 2030 Strategy, CAREC will develop its climate actions based on five principles:

(i) **aligning** with national strategies and with the United Nations Framework Convention on Climate Change's Paris Agreement;
(ii) **deepening** regional cooperation for climate change mitigation and adaptation actions;
(iii) **expanding** coordination with all development partners on regional climate action;
(iv) **integrating** the role of the private sector and civil society into the regional climate dialogue; and
(v) **building** an open and inclusive regional platform on climate change.

Mongolia. Coal briquettes are packed and distributed around the *ger* district in Ulaanbaatar. Compared to raw coal, the briquettes emit less smoke and ash (photo by Ariel Javellana/ADB).

3 Implications of Climate Change for Regional Cooperation

Climate change presents a major challenge for CAREC countries. The effects of climate change are already evident. Particularly notable in the CAREC region are higher than average increases and greater variability in temperature and extremes in weather and precipitation, resulting in greater water scarcity; more floods, landslides, wildfires, and droughts; the melting of glaciers; expanding desertification; declining agricultural productivity; food insecurity; migration; worsening health outcomes; and conflict. The years 2022–2023 have witnessed particularly dramatic and deadly examples of the impacts of climate change in the region, including the devastating floods in Pakistan, severe drought in Afghanistan, drought and floods in the PRC, weeks of excessive heat beyond historical levels, and transboundary water issues in Central Asia. These trends and climate-linked events are sharp reminders of the long-term prospects of even more serious impacts of climate change if steps are not taken to control carbon emissions and increase countries' resilience.

Climate change in the CAREC region requires urgent and collective action. To mitigate carbon emissions and improve resilience to climate change, CAREC countries will have to pursue an inevitable energy transition (with greater energy efficiency and investments in renewable energy investment), more efficient use of water, climate-smart agriculture and industry, and green heating and cooling technologies.[5] The countries will also have to invest in much greater resilience of their economies, with climate-responsive design of transport and cities; more effective health systems; improved risk management; early warning systems to mitigate climate-linked disasters; and strengthened education, social security, and financial management.

Climate change has regional impacts, and many climate issues need to be addressed on a regional basis for maximum effect. As climate change affects weather and climate conditions beyond borders, it requires strengthening data collection and analysis; creating regional centers of weather and climate observations and prediction; and implementing coordinated risk management, planning, and action. Investment projects and policies related to energy, transport, water, agriculture, urban infrastructure, disaster risk management, and financing have important regional spillovers and, hence, must be climate-informed and climate-coordinated in planning and implementation. Moreover, a regional approach to technology transfer and knowledge sharing, research and data, analytic risk modeling, and capacity building in these areas, as well as in education, health and social protection, results in a regional public good through the creation and diffusion of relevant technology, knowledge, and best practices. Regional climate action requires a readiness by countries to cooperate, a coordinated strategy that complements national strategies, and institutional capacities to implement such strategies.

---

[5] CAREC Institute. 2022. *Post-Pandemic Framework for a Green, Sustainable and Inclusive Recovery. Background Report*.

Tajikistan. Embankments along the Pyanj River in Hamadoni District prevent flooding from mountain snowmelt, a recurring problem expected to worsen as the climate warms (photo by Nozim Kalandarov/ADB).

4 Central Asia Regional Economic Cooperation: A Regional Platform for Climate Action

CAREC can play a leading role in supporting coordinated climate actions among member countries. CAREC is an important regional convenor, as it offers great opportunities to operationalize climate agenda in its five operational clusters and promote developing member countries' (DMCs) and development partners' active and sustained participation at policy and project levels, including supporting crosscutting priorities. Given the regional interconnectedness of CAREC countries in many climate issues, the strong commitment of the countries to work together and use the CAREC platform to identify linkages and possible regional solutions for climate change issues are needed in the operational clusters providing overall cross-cluster guidance, coordination, and monitoring.

Development partners shall strongly support the implementation of the CAREC Climate Change Vision. CAREC will continue to serve as a platform for cooperation and coordination among development partners in support of DMCs' regional climate action. Development partners such as the Asian Development Bank (ADB), Asian Infrastructure Investment Bank (AIIB), European Bank for Reconstruction and Development, International Monetary Fund, Islamic Development Bank, United Nations Development Programme, and the World Bank have been supporting CAREC in different areas directly or through their regional initiatives and have contributed critical financing for projects of regional significance along with technical assistance, capacity-building, research and coordination (Appendix 1). Under the CAREC 2030 Strategy, all potential development partners are invited to join CAREC activities by encouraging their active participation in all sectors and thematic areas and coordinating their plans and actions. This well-functioning mechanism allows CAREC to serve as a coordinating body for development partners' engagement in regional climate actions through collaboration in cluster activities, participation in the climate change institutions (see Chapter 7), and through development partners' Forum and related events.[6] The multilateral development banks (MDBs) will continue to play a major role in supporting CAREC. In June 2023, MDBs signed the joint principles on the alignment of their new operations with the goals of the Paris Agreement.[7] These principles are also expected to govern their engagement in support of CAREC's climate agenda.

---

[6] CAREC could become a platform for developing and testing a regional approach under a Just Energy Transition Partnership (JETP) for the CAREC region. JETP is a "financing cooperation mechanism, the aims of which are to help a selection of heavily coal-dependent emerging economies make a just energy transition. The goal is to support these countries' self-defined pathways as they move away from coal production and consumption while doing so in a way that addresses the social consequences involved, such as by ensuring training and alternative job creation for affected workers and new economic opportunities for affected communities." (https://www.iisd.org/articles/insight/just-energy-transition-partnerships). (https://www.adb.org/news/features/update-energy-transition-mechanism-april-2023).

[7] MDBs Agree Principles for Aligning Financial Flows with the Paris Agreement Goals, ADB News Release, 20 June 2023. https://www.adb.org/news/mdbs-agree-principles-aligning-financial-flows-paris-agreement-goals. The MDB signatories to these principles are the African Development Bank, ADB, AIIB, Council of Europe Development Bank, European Bank for Reconstruction and Development, European Investment Bank, Inter-American Development Bank Group, Islamic Development Bank, New Development Bank, and the World Bank Group.

The CAREC Institute can drive a strong knowledge base and capacity building in sector and thematic areas with a regional focus on climate change. As documented in the 2022 CAREC Climate Change Scoping Study (footnote 2), CAREC and the CAREC Institute have been engaged in important climate change knowledge work. This knowledge and capacity-building work will be intensified in close partnership with academic institutions and think tanks in the DMCs and with development partners.

CAREC will explicitly include climate action as a crosscutting priority in its 2030 Strategy. The Strategy, as approved in 2017, referred to climate change as a regional challenge and has broadly indicated that it will support regional actions that must complement national efforts to address the United Nations' Sustainable Development Goals and climate change. It is timely and crucial to elevate climate actions as a crosscutting thematic area, and by pursuing the climate agenda throughout the various clusters, CAREC shall bring together different actors to find the best possible financial and institutional solutions. Given the urgency of regional climate action and its crosscutting nature, climate actions should be promoted in the CAREC 2030 Strategy as a crosscutting priority along with gender and information and communication technology. The mid-term review in 2024 will be an opportunity to evaluate the 2030 Strategy and how to make it relevant and responsive to the emerging climate-related needs and priorities to build a sustainable and climate change-resilient future for the CAREC countries and beyond.

5 Central Asia Regional Economic Cooperation's Focus on Climate Change in Strategies and Programs

CAREC shall explore opportunities to support its DMCs in designing and implementing their national climate strategies with a special focus on regional dimensions. CAREC shall assist DMCs in assessing, refining, and implementing their nationally determined contributions (NDCs), national adaptation plans, and other climate-related priorities. A preliminary review of the NDCs of the 10 CAREC member countries (i.e., Azerbaijan, the People's Republic of China, Georgia, Kazakhstan, Kyrgyz Republic, Mongolia, Pakistan, Tajikistan, Turkmenistan, and Uzbekistan) shows different levels of ambitions and depth of information and implementation plans. Further analysis, benchmarking, tracking, and support for the implementation of NDCs on a regional basis, are needed. CAREC shall explore opportunities to work on NDCs and national adaptation plans to identify potential gaps and opportunities for regional cooperation and common policy interest.

CAREC shall review and amend, as needed, the sectoral and thematic strategic documents in its five operational clusters, as well as the program's project classification methodology, to align with the Paris Agreement and MDB joint principles. Currently, CAREC has only one policy document, i.e., the CAREC Health Strategy 2030, that reflects the criteria of the MDB joint principles. The CAREC Water Pillar Scoping Study also refers to the MDB joint principles. The strategic documents in other sectors and thematic areas may need to be revisited, to explicitly incorporate the Paris Agreement and the MDB joint principles. In addition, the CAREC project classification methodology must include climate change in the project classification for better monitoring and reporting.

CAREC shall pursue its regional climate agenda by focusing on seven priority areas: (i) energy; (ii) water; (iii) agriculture; (iv) transport, transit, and trade; (v) climate-smart cities; (vi) climate-related disasters; and (vii) health, education, and social protection. The agenda is ambitious and will be implemented in line with DMC demands and with CAREC's ability to manage the various strands of potential activities. Much will depend on how effectively the cluster committees and expert groups manage to incorporate climate agenda into their strategies, analysis, dialogue, and actions.

Energy

CAREC shall explore opportunities to support a just and effective transition to a low-carbon energy system. CAREC DMCs still face challenges to ensuring universal energy access, safeguarding energy security, improving energy sector governance; and regional energy trade, and enhancing sustainability. Despite progress in some countries to achieve national energy security, CAREC DMCs still face significant gaps in meeting future energy demand while managing the transition to renewable energy sources. Greater reliance on the region's hydro, solar, and wind resources needs to be underpinned by effective interconnectivity for regional electricity trade. CAREC will revisit and implement its Energy Strategy 2030 to support cross-border investments, technological transfer, development of planning and implementation capacity, and, most importantly, the development of effective national policy and regulatory conditions that are Paris Agreement aligned and mutually beneficial. CAREC shall continue to play a key role in this process.

ADB, in cooperation with other development partners, will further explore ways to deploy the energy transition mechanism (ETM) in the CAREC region. ETM accelerates the retirement of existing coal-fired power plants and combined heat and power coal-fired plants and pursues strategic decarbonization of the power, construction, and industrial sectors. Kazakhstan and Pakistan are finalizing a high-level assessment of existing assets suitable for ETM. CAREC will further explore the opportunity to support DMCs in developing skills needed to ensure a smooth and just transition to green jobs and access to finance to participate in the low-carbon transition. CAREC will continue to support DMCs to develop adaptive and responsive social protection mechanisms to protect the poor and vulnerable, women and girls, persons with disabilities, as well as elderly and young people during the transition.

Water

CAREC shall further explore the opportunities to support DMCs' national and regional adaptation action in the water sector. Water is the lifeblood of development in the CAREC region for agriculture, industry, and human consumption. Much of the CAREC region suffers from perennial water scarcity, aggravated by the highly unequal distribution of water resources across the country and region, high population growth, poor water policies, inefficient water use, and by increasing climate change effects. These include variability of precipitation, melting of glaciers, and highly variable river flows, with alternating droughts and floods leading to major economic and social dislocations. The almost total disappearance of the Aral Sea over the last 5 decades demonstrates the dramatic impact of poor water management on long-term environmental and climate outcomes and, hence, on regional communities. In line with the Resolution of the United Nations General Assembly dated 14 December 2022 (77/158) to declare 2025 the International Year of Glaciers' Preservation and following the 2021 CAREC Water Pillar Scoping Study, CAREC shall further explore opportunities to assist its DMCs in addressing issues related to accelerated melting of glaciers and its consequences; protecting regional water resources by fairly and efficiently sharing and utilizing them nationally and regionally, including with improved water storage and management systems; and identification of appropriate investments and technical support that are based on sound climate risk vulnerability assessment.

Agriculture

CAREC member countries shall further explore opportunities to support the development of a climate-smart agriculture system in the region. Agriculture is an important sector in the CAREC region with a critical role in economic growth, employment, poverty reduction, food security, and climate mitigation and adaptation. Under CAREC's Cooperation Framework for Agricultural Development and Food Security, agricultural policies and practices need to adapt to the impending negative impact of climate change on agricultural productivity and food security and contribute to mitigation efforts.[8] Based on its strong engagement in the agriculture and water cluster, CAREC will assist with adaptation and mitigation in the agricultural sector by (i) improving the productivity of agriculture through climate-smart agricultural policies, technologies, and methods, based on, wherever possible, the transfer of green technologies and best policies and practices; (ii) changing farming and livestock practices to reduce GHG emissions and husbandry; (iii) encouraging innovation and digitalization technologies, development of food value chains, diversification of crops, and the introduction of drought-resistant crops; (iv) improving regional management of water resources, including more efficient irrigation systems and the Aral Sea restoration program; (v) strengthening forestry management and sustainable development, including afforestation and reforestation; and (vi) supporting the development of an efficient trading system for agricultural inputs and outputs throughout the region.

Transport, Transit, and Trade

CAREC shall explore opportunities to assist its DMCs in reducing the carbon footprint of regional transport services (road, rail, and air) while improving regional connectivity with climate-smart and efficient transport technologies, regulations, and border-crossing policies and infrastructure. Regional transport and corridor development, along with transit and trade facilitation, has long been a CAREC priority since transport is critical for national and international connectivity for many landlocked countries, allowing improved movement of people and goods and better links to global value chains. However, transport is also one of the most carbon-emitting and polluting sectors, and transport assets are vulnerable to the impacts of climate change. CAREC shall strengthen the CAREC Transport Strategy 2030's alignment with the Paris Agreement, including through support for green freight, urban and intercity transport electrification, economic corridor development, direct air transportation connections, and climate-smart transport policies to maximize the climate resilience of new infrastructure and deliver low-carbon infrastructure in the countries. CAREC will embark on the Decarbonization of the Transport Sector study, which will assess existing and future energy consumption in the transport sector and ascertain how different policies could change future levels.

CAREC will promote the greening of regional trade and explore how trade can be part of climate solutions. The CAREC Integrated Trade Agenda 2030's midterm review in 2024 will provide an opportunity to help DMCs understand the impact of the carbon footprint of trade on their economies, including carbon tracking and reporting across global supply chains. CAREC shall examine and develop policy advice in exploring country-specific and regional responses to decarbonization measures such as the European Union's (EU) Carbon Border Adjustment Mechanism to facilitate CAREC countries' sustainable trade with their trading countries and enhance compliance. The CAREC Integrated Trade Agenda 2030 and its institutional mechanisms will explore areas of cooperation and regional approaches for trade to be part of climate change solutions. To support this, a scoping study will be undertaken based on (i) increasing access and incentivizing trade of environmental goods; (ii) mainstreaming multilateral climate agreements in regional arrangements; (iii) digitizing trade processes and promoting cross-border paperless trade; (iv) investing in smart and resilient border infrastructure facilities; and (v) accelerating diversification to reduce heavy reliance on trade or use of fossil fuels. Under the trade, tourism, and economic corridor cluster, CAREC will assess how improvements in trade policy, trade facilitation

---

[8]   CAREC. 2022. *Cooperation Framework for Agriculture and Food Security in the CAREC Region.* https://www.carecprogram.org/uploads/CAREC_MC_2022_2a_Agriculture-Food-Security-Framework-EN.pdf.

and transit management, logistics, services, and tourism development will contribute to climate mitigation and adjustment.

Cities

CAREC shall support the development of climate-smart cities. Cities—and, more generally, urban areas—are key to meeting climate mitigation targets, adaptation goals, and resilient growth objectives. Urbanization has been progressing rapidly in the CAREC region and will continue. Urban centers are the hub of economic activity and are traditionally carbon-intensive due to industrial activity, urban transport, waste, and the heating and cooling needs of the population. Many of the traditional policy recommendations for greater urban efficiency are climate-smart as they promote carbon emissions reduction, denser residential development planning rather than urban sprawl, sound zoning, building codes, reduced reliance on the private automobile in exchange for mass transit, bicycling and walking, and efficient provision of municipal services to all (electricity, water and sewerage, garbage disposal, central home heating, etc.). However, much more can be done, with a specific focus on climate mitigation and adaptation. These include support for sustainable transport that relies on renewable energy (including electrified mass transit and private automobiles), low-carbon district and individual home heating, more efficient cooling, and increased reliance on "circular economy" measures (including recycling of plastic, batteries, electronic equipment, etc.).

Climate-smart city policies are nationally and locally driven, but regional approaches can help with design and implementation. Regional initiatives and city alliances among neighboring countries can support national and local policy actions by sharing experience, building capacity, supporting technology transfer, harmonization of building standards, benchmarking and monitoring progress, and building coalitions that create greater public awareness and policy commitment. CAREC had good examples of regional support that demonstrated cooperation in promoting low-carbon development in selected CAREC cities in Kazakhstan, Mongolia, and the PRC.[9] CAREC's support for the Almaty–Bishkek Economic Corridor demonstrates how improved regional connectivity between the cities can help achieve the goal of more livable, efficient, and resilient cities.

Disaster Risk Management

CAREC shall assist countries in promoting climate and disaster resilience. The CAREC region is highly exposed and vulnerable to natural hazards, including those caused by extreme weather and climate events. The frequency and intensity of such disaster events are expected to increase, creating major adaptation challenges. Disaster events often transcend national borders and create significant fiscal challenges for individual countries. Even when the effects are localized within a country, early warning requires weather and climate observations in neighboring countries. Moreover, cooperation among CAREC countries can lead to valuable data and knowledge exchange and will significantly reduce the cost of disaster risk financing for immediate emergency relief, rehabilitation, and reconstruction once a disaster event occurs.

---

[9]   See ADB. 2022. *Promoting Low-Carbon Development in Central Asia Regional Economic Cooperation Program Cities.* https://www.adb.org/sites/default/files/project-documents/50287/50287-001-tcr-en.pdf.

CAREC can serve as a platform to support complementary regional disaster risk management and disaster risk financing initiatives. ADB is supporting comprehensive disaster risk financing solutions in the CAREC region, and tools to improve countries' disaster risk modeling capacity through quantifying the benefits and the costs of climate and disaster risk reduction and finance measures adopting a risk-layered approach. Regional and national disaster risk finance initiatives are complementary and include the implementation of climate adaptation and disaster risk reduction measures. Regional initiatives—such as the establishment of a disaster risk transfer facility in the long term and the issuance of disaster relief bonds to provide immediate (non-debt) financing for emergency relief and rehabilitation in the short term—have the potential to bring together technical and financial resources from governments, development partners, and the private sector to build macroeconomic stability and enhance the physical and financial resilience of the most vulnerable. The international community is also focusing on early warning capacities in vulnerable countries through the Early Warnings for All Initiative of the UN Secretary-General.[10]

Social Development Actions: Health, Education, and Social Protection

Under its human development cluster, CAREC shall explore further opportunities to support DMCs in developing regional climate-related actions in the areas of health, education, and social protection. While regional linkages and spillovers in these areas are less prevalent than in the other priority areas, they do matter. Regionally coordinated action is of importance in terms of preparedness, capacity building, policy formulation and implementation, and the sharing of data, research, lessons, and monitoring. Climate change has a significant impact on health conditions in the CAREC region through the spread of disease, lack of safe drinking water, reduced nutrition, excessive heat, or life-threatening floods. Improvements in health systems will be critical to support effective adaptation. For education, the development of advanced skills in climate research and management through appropriate tertiary and technical education on a regional basis will be helpful. General education and increased awareness of climate change trends, causes, impacts, and policies, suitably tailored to primary, secondary, and adult education curricula, are also important. Social protection systems are essential to respond quickly in the wake of disaster events and to deal with the potentially negative impacts of energy transition measures especially increases in energy prices. CAREC shall implement and share DMCs' approaches and lessons in the design and implementation of social protection measures.

Other Areas for Possible Central Asia Regional Economic Cooperation Climate Engagement

The CAREC Secretariat and the CAREC Institute shall further explore other potential areas of engagement. Specifically, they will coordinate analytical and capacity building activities to help DMCs better understand climate and disaster risk and support risk-informed policies and investments and look into emerging issues and potential regional actions identified in the 2022 CAREC Climate Change Scoping Study, including:

　(i)　**Macroeconomic policy coordination.** Climate change will raise several important macroeconomic policy issues, including fiscal, green procurement and debt sustainability,

---

[10]　World Meteorological Organization (WMO). *Early Warnings for All initiative,* https://public.wmo.int/en/earlywarningsforall.

inflation control, tax, and public expenditure management. Coordination of policies across CAREC countries and mutual monitoring, sharing of lessons and support in a macroeconomic crisis are important areas for CAREC engagement. The International Monetary Fund could be the appropriate DP to take a lead in this area.

(ii) **Emissions Trading System (ETS).** ETS is potentially an effective and efficient way of reducing carbon emissions. Some CAREC countries have already introduced such a system. CAREC could commission the CAREC Institute to review the design and experience in other countries worldwide and advise member countries on better national ETS and study the possibility of accessing international carbon markets.

(iii) **Nature-based climate solutions (NBS).** NBS are actions to protect, better manage, and restore nature to reduce GHG emissions, store carbon, and enhance climate resilience. CAREC has large potential for employing cost-effective NBS in different sectors. NBS are cheaper and more flexible than conventional alternatives for addressing climate risk. For example, they can contribute to rural development, sustainable livelihoods, and food security by adopting a complete-food-chain approach to agriculture and protection against land degradation that, among other benefits, boosts agricultural yields. In the urban context, NBS can prevent flooding, while also creating green spaces to make cities more livable. They an also advance gender equality by promoting economic empowerment and providing women with income-generating opportunities. Various MDBs support NBS in CAREC DMCs and could share their findings in a systematic manner with the support of the CAREC Secretariat.

(iv) **Green public sector management.** Regional cooperation within the public sector presents a promising avenue for tackling climate change challenges collectively. By fostering shared learning and knowledge exchange in the areas of carbon pricing and fossil fuel subsidy reform, green public finance management, green procurement, and state-owned enterprise reform for greening of investment, countries can draw from each other's experience and best practice to implement more effective climate policies. Regionally leveraging pooled resources, both financial and technical, can enhance the efficiency and impact of climate initiatives.

6 Financing Climate Action

CAREC shall prioritize the mobilization of financial resources and mechanisms, by exploring opportunities to support DMCs in the development of financing plans and innovative instruments for regional climate action programs in close partnership with development partners and private sector investors. Climate mitigation and adaptation will require large financial resources. CAREC will assist DMCs in developing their climate finance plans and instruments, including domestic resource mobilization as well as access to external official finance and catalyzing capital markets and private sector investors. This requires increased coordination and collaboration among development partners to pool financial resources, share and transfer risks, and combine knowledge and technical expertise in planning and implementing climate-informed development programs or projects. CAREC shall continue to work with DMCs to strengthen their project preparatory capacities and catalyze financing for climate change by establishing a dedicated financing mechanism such as the CAREC Climate and Sustainability Project Preparatory Fund. CAREC will assist DMCs in accessing international climate funds, in particular the Green Climate Fund, the Global Environment Facility, the Adaptation Fund, the Climate Investment Funds, and the Systematic Observations Financing Facility, and will support cooperation under the frameworks of South–South cooperation and the Belt and Road Initiative. Moreover, CAREC will provide capacity building for attracting climate financing and for developing and using innovative disaster risk financing instruments for DMCs. Throughout, CAREC will continue its efforts to scale up effective regional climate solutions in the pursuit of a sustainable development pathway featuring green, low-carbon, and resilient development.

Mongolia. Herders promote climate-smart agricultural activities, cultivating fodder or animal feed that is more resilient to extreme weather changes, and using plants that adapt to droughts (photo by Eric Sales/ADB).

7 Guiding and Monitoring Results for Implementing the Climate Change Vision

CAREC shall establish a high-level Climate Change Steering Committee. This will consist of government officials represented at the national focal point level, or higher, as proposed by the CAREC authorities. It will be responsible for (i) overseeing and guiding CAREC's efforts on climate action; (ii) coordinating and providing guidance to relevant CAREC institutions and sector committees and working groups in the review of the strategies, action plans, and project portfolios of the five priority clusters and crosscutting priorities to ensure alignment with and support for the implementation of the CAREC Climate Change Vision; and (iii) developing a 3-year rolling CAREC Climate Change Action Plan and monitoring its implementation. The Committee will report annually to the CAREC Ministerial Conference on the progress of the CAREC Climate Action Plan. Appendix 2 contains draft terms of reference for the Committee. The CAREC Secretariat will, in consultation with national focal points, develop final terms of reference for the Committee and recommendations on the level and staffing for review and approval by the Senior Officials' Meeting.

The CAREC Climate Change Steering Committee will be assisted by a Cross-Sectoral Working Group on Climate Change and/or a Climate Expert Group. The Cross-Sectoral Working Group shall focus on the Paris Agreement alignment of the activities in the CAREC priority clusters and crosscutting priorities. The Expert Group brings together climate and finance experts from the CAREC region and internationally to advise the Steering Committee on the technical aspects of the CAREC climate agenda.

Tajikistan. In an effort to reduce pollution, a cyclist bikes in the Tajik–Uzbek border along the Ayni–Panjakent road. The Central Asia Regional Economic Cooperation Corridor 6 (Ayni–Uzbekistan Border Road) Improvement Project will restore approximately 113 kilometers of road between Ayni–Panjakent in the northern part of the country and its border with Uzbekistan (photo by Nozim Kalandarov/ADB).

8 Summary of Recommendations for Ministers

The following recommendations are submitted to CAREC Ministers for consideration and endorsement:

- Ministers to endorse the CAREC Climate Change Vision document including the Vision statement: *"A Region of Sustainable Development, Shared Prosperity, and Climate Resilience."*
- CAREC to explicitly include climate change as a crosscutting priority in the CAREC 2030 Strategy.
- CAREC to establish the CAREC Climate Change Steering Committee, which could be advised by a Cross-Sectoral Working Group on Climate Change and/or a Climate Expert Group.
- The CAREC Climate Change Steering Committee to develop a mid-term CAREC Climate Change Action Plan, which will prioritize and phase in the actions recommended for consideration in this vision document.
- The CAREC Secretariat and CAREC Institute to coordinate their respective analytical and capacity building activities to help DMCs better understand climate and disaster risk and support risk-informed policies and investments.
- The CAREC Secretariat to organize CAREC Development Partners Forums to strengthen cooperation and create synergies of supported initiatives to operationalize the Climate Change Vision document.
- The CAREC Secretariat to develop a communication and outreach plan for the CAREC Climate Change Vision document.

Kazakhstan. Sunset over traffic on the bypass road (photo by Igor Burgandinov/ADB).

Appendixes

Development Partners' Climate-Related Support to the Central Asia Regional Economic Cooperation Program

The Asian Development Bank (ADB) is supporting the climate agenda by drawing on a wide range of its Asia-Pacific-wide climate initiatives. ADB has made major changes and driven key innovations in becoming the Climate Bank for Asia and the Pacific. ADB has increased its commitment to providing $80 billion of its own funds to climate finance between 2019 and 2030 to the ambition of $100 billion through the development of a pipeline of quality projects and programs. In 2023, ADB set up a new Innovative Finance Facility for Climate in Asia and the Pacific, a landmark program that aimed to significantly ramp up support to developing member countries (DMCs), including CAREC DMCs, in the battle against climate change. It also created the Energy Transition Mechanism and Just Transition to provide financial structure for public and private sectors incentives to facilitate early retirement of coal-based power and combined heat and power plants.

ADB is also supporting the creation of the CAREC Climate and Sustainability Project Preparatory Fund (to be launched in 2024) to secure essential financing for project preparatory and new initiatives in climate change for CAREC DMCs. ADB will improve its intervention in renewable energy projects through the active participation of the private sector. ADB is also working closely with CAREC countries to design and implement regional disaster risk financing solutions that can complement national climate adaptation and disaster risk management efforts. All these initiatives provide potential financial and technical resources which ADB can put to work in support of regional climate action in the CAREC region.

Asian Infrastructure Investment Bank (AIIB) has adopted connectivity and regional cooperation as one of the thematic priorities of AIIB's Corporate Strategy charting its path to 2030. Connectivity holds one of the keys to address the climate crisis by forming the regional and/or sectoral approaches and perspective, as recognized in AIIB's first ever Climate Action Plan (CAP) launched in September 2023. The CAP articulates AIIB's ambitions and guides its climate solutions to support its members and clients by (i) providing tailored solutions to meet Members' needs; (ii) taking a holistic approach to climate mitigation, adaptation, resilience, and biodiversity; (iii) scaling climate finance through catalytic mobilization; and (iv) promoting technological innovation. AIIB stands ready to work with CAREC countries and partners to make a region of sustainable development, shared prosperity, and climate resilience, including helping to create a fluid marketplace, where investment concepts and proposals are matched with appropriate financing for project preparation, development, and implementation from across development partners.

European Bank for Reconstruction and Development (EBRD). Within the CAREC region, EBRD operates in Kazakhstan, Kyrgyz Republic, Tajikistan, Turkmenistan, and Uzbekistan. It supports decarbonization and strengthening of climate resilience in key economic sectors through providing various climate policy support, including supporting countries develop low-carbon and climate-resilient pathways (LCPs) at national, sectoral, and subnational levels. In Kazakhstan and Uzbekistan, EBRD has supported the development of LCPs in the power, industry, and agri-food sectors. To implement the Global Methane Pledge, EBRD is engaged in Kazakhstan and Uzbekistan with respect to investments in gas, waste, and agricultural sectors. These LCPs work under the Global Methane Pledge and feed into the development of Paris-aligned long-term strategies, helping inform decisions regarding net-zero and adaptation commitments at both sectoral and economy-wide levels. EBRD also engages with government ministries and regulators to promote market liberalization in key areas, such as strengthening regulatory capacity in electricity markets and the water sector, developing tariff methodologies, and improving governance and tariffs. These policy activities are closely linked to investments EBRD undertakes in relevant sectors, such as in energy generation and distribution companies, municipal infrastructure, and irrigation/water infrastructure. EBRD has supported the development of the regulatory framework for competitive procurement of renewable energy projects in Kazakhstan and Uzbekistan, which has led to successful solar and wind auctions. EBRD also has a track record on engagement with authorities of Kazakhstan in supporting the transition to cleaner fuels through the domestic emissions trading system and developing the supply chain for low-carbon fuels, such as green hydrogen, green ammonia, and sustainable aviation fuels. In the Kyrgyz Republic, EBRD is in the final stages of supporting the development of a roadmap for utility-scale renewables development through private sector investment. Since 2018, it has been supporting the Climate Finance Center in the Kyrgyz Republic and its project preparation facility, to facilitate the nation's green development priorities. At the city level, EBRD implements its flagship Green Cities program to help cities identify, benchmark, prioritize and invest in Green City measures to improve urban environmental performance. This includes delivering strategy and policy support, facilitating, and stimulating Green City infrastructure investments, building the capacity of city administrators and key stakeholders, and supporting access to green finance. EBRD also promotes greening of broader financial systems within countries through supporting regulators and partner financial institutions to introduce climate corporate governance practices, in line with Task Force on Climate-Related Financial Disclosures recommendations, providing green financial products, such as EBRD's flagship Green Economy Financing Facilities, and supporting host countries in establishing the Chapter Zero initiatives, a capacity building and knowledge-sharing platform that builds a community of non-executive directors to lead boardroom discussions on climate change impacts, and help companies develop and implement robust transition plans and measurable actions.

Islamic Development Bank (IsDB) places great importance on addressing climate change in its member countries and invests in projects that have a high impact in various sectors such as renewable energy, smart systems, climate-smart agriculture, forestry, resource efficiency, zero-carbon buildings, sustainable transport, and sustainable cities. IsDB has increased its efforts to ensure that all its operations and activities in CAREC countries align with the Paris Agreement goals by systematically integrating climate mitigation opportunities and adaptation measures in its investments and lending operations. IsDB has set a target to achieve a climate finance target of 35% by 2025, guided by its

Climate Action Plan 2020–2025. This plan is designed to support the member countries' climate action goals, including those of CAREC countries. IsDB is collaborating closely with CAREC countries to create and carry out investment projects that will help them attain their Nationally Determined Contributions. These targets are also included in the IsDB Member Countries Strategic Partnership and Country Engagement Framework of CAREC countries. Specifically, IsDB is working with Tajikistan to plan and execute the construction of 3,600 megawatt (MW) Rogun hydropower plant valued at $150 million. IsDB supports the Tajikistan government in pursuing a green economy to improve economic resilience and promote sustainable and efficient development while considering climate change, and environmental and socio-economic challenges. The proposed project would offer economic advantages by exporting electricity to Central Asian countries such as Uzbekistan and Kazakhstan, which have a combined demand of over 62,000 gigawatt hours and 110,000 gigawatt hours, respectively, as of 2020.

World Bank is committed to mainstreaming climate change into its Central Asia country portfolio of $12 billion and supporting CAREC's climate goal of better cooperation through analytics, support for policy reform, and investment programs. Some of the key initiatives on climate change include the following:

(i)   Country Climate and Development Reports (CCDRs) are core World Bank analytic products that assess the interplay of climate change and development in the country, identify key policy reforms and investment needs to shift country development paths to be climate-resilient and low-carbon, and inform how climate considerations can be mainstreamed into World Bank portfolios and support climate action in Central Asian countries.

(ii)   The Central Asia Water and Energy Program, operational since 2009, is a joint effort supported by the European Union, Switzerland, the United Kingdom, United States Agency for International Development (at early phases) and managed by the World Bank. Since the start of the program, it has facilitated 18 water, energy, and environmental investments worth approximately $2.6 billion. The fourth stage of the program to be implemented during 2024–2028 would be strategically oriented toward promoting regional cooperation for more resilient and better integrated water and energy management under a changing climate.

(iii)   The Central Asia Resilient Landscape Restoration Program is a major regional effort to mitigate climate change that are both caused by and are leading to land degradation, deforestation, and vulnerability of ecosystems. The program (total program budget of $257 million) supported by ongoing projects in Tajikistan, Uzbekistan, and Kazakhstan and a planned operation in Kyrgyz Republic, will increase resilience of regional landscapes in Central Asia through landscape restoration.

(iv)   The World Bank offers comprehensive support for clean energy transitions in Central Asia through upstream and financing critical investments, including large-small hydropower facilities (Kyrgyz Republic, Tajikistan), large-scale deployment of solar projects (pilot and scale-up in Uzbekistan, [up to 2,500 MW], Kyrgyz Republic [up to 500 MW] and Tajikistan [200 MW]). In addition, the World Bank is assisting the countries in establishing and operationalizing carbon markets, including through the global-first Innovative Carbon Resource Application for Energy Transition international carbon trade project in Uzbekistan.

(v)   The Coalition for Climate Change Action Program aims to support regional knowledge sharing and capacity building activities on climate change and climate action specifically targeted at ministries of finance. The program is at its core a country- and Ministry of Finance-driven initiative that could support CAREC countries' vision on climate in those areas.

APPENDIX 2

Draft Terms of Reference for the Central Asia Regional Economic Cooperation Program Climate Change Steering Committee

- Oversee and guide CAREC's framework on climate action.
- Coordinate with and provide guidance to other CAREC institutions and sector-specific committees and working groups in the review of the strategies, action plans, and project portfolios of the five priority clusters and crosscutting priorities to ensure alignment with and support for the implementation of the CAREC Climate Change Vision.
- Develop the CAREC Climate Change Action Plan.
- Develop the CAREC climate project portfolio.
- Launch specific initiatives to strengthen the enabling environment for climate action by harmonizing the legal and regulatory environment and by building capacity.
- Build and maintain a multistakeholder consensus across the region.
- Gather and share best practices for regional climate action.
- Develop and launch a strategic communications plan for CAREC's Climate Change Action Plan.
- Work with development partners to secure project funding.
- Build public-private partnerships for regional climate action.
- Establish a monitoring system to measure and annually report to the CAREC Ministerial Conference on progress with the implementation of the CAREC Climate Change Action Plan.
- Review and update the CAREC Climate Change Vision document and the CAREC Climate Change Action Plan every 3 years to adapt to changing conditions and needs.

Glossary

Adaptation	In human systems, adaptation refers to the process of adjustment to actual or expected climate and its effects to moderate harm or exploit beneficial opportunities. In natural systems, adaptation refers to the process of adjustment to actual climate and its effects; human intervention may facilitate adjustment to expected climate and its effects.
Carbon pricing	Refers to an approach that internalizes the external cost of damage caused by climate change, by putting a price on greenhouse gas emissions.
Climate change	Long-term shifts in temperatures and weather patterns due to natural processes and human activities.
Disaster	A serious disruption of the functioning of a community or a society triggered by geophysical or extreme weather hazard events leading to human, material, economic, or environmental losses and impacts. Disasters occur when hazard events and extreme weather hazard events interact with the exposure of vulnerable people and assets to those events.
Energy transition mechanism	A scalable, collaborative initiative that will leverage a market-based approach to accelerate the transition from fossil fuels to clean energy.
National adaptation plans	The objectives of national adaptation plans are to (i) reduce vulnerability to the impacts of climate change by building adaptive capacity and resilience; and (ii) integrate adaptation into new and existing national, sectoral, and subnational policies and programs, especially development strategies, plans and budgets.
Nationally determined contributions	The efforts that countries around the world pledge to reduce their greenhouse gas (GHG) emissions and adapt to the impacts of climate change.
Nature-based solutions	Actions to protect, sustainably manage, or restore natural ecosystems that address societal challenges such as climate change, human health, food and water security, and disaster risk reduction effectively and adaptively, simultaneously providing human well-being and biodiversity benefits.

continued on next page

Glossary *continued*

Net zero	Cutting GHG emissions to as close to zero as possible, with any remaining emissions re-absorbed from the atmosphere through natural processes or carbon capture technologies.
Paris Agreement	A legally binding international treaty on climate change. It was adopted by 196 parties at the UN Climate Change Conference (COP21) in 2015. Its overarching goal is to hold "the increase in the global average temperature to well below 2°C above pre-industrial levels" and pursue efforts "to limit the temperature increase to 1.5°C above pre-industrial levels".
Resilience	The capacity of interconnected social, economic, and ecological systems to cope with a hazardous event, trend, or disturbance, responding or reorganizing in ways that maintain their essential function, identity, and structure, while maintaining a capacity for adaptation, learning, and/or transformation.

Source: Asian Development Bank.

References

ADB. 2023. *CAREC 2030: Supporting Regional Actions to Address Climate Change–A Scoping Study*. https://www.adb.org/sites/default/files/publication/879296/carec-2030-regional-actions-climate-change.pdf.

ADB. 2023. *Evaluation of ADB Support for the Central Asia Regional Economic Cooperation Program, 2011–2022*. https://www.adb.org/documents/evaluation-adb-support-central-asia-regional-economic-cooperation-program-2011-2022.

ADB. 2023. MDBs Agree Principles for Aligning Financial Flows with the Paris Agreement Goals. ADB News Release. 20 June 2023. https://www.adb.org/news/mdbs-agree-principles-aligning-financial-flows-paris-agreement-goals.

ADB. 2023. *Update on ADB's Energy Transition Mechanism–April 2023*. https://www.adb.org/news/features/update-energy-transition-mechanism-april-2023.

ADB. 2022. *Promoting Low-Carbon Development in Central Asia Regional Economic Cooperation Program Cities*. https://www.adb.org/sites/default/files/project-documents/50287/50287-001-tcr-en.pdf.

CAREC. 2022. *Cooperation Framework for Agriculture and Food Security in the CAREC Region*. https://www.carecprogram.org/uploads/CAREC_MC_2022_2a_Agriculture-Food-Security-Framework-EN.pdf.

CAREC Institute. 2022. *Post-Pandemic Framework for a Green, Sustainable and Inclusive Recovery: Background Report*. https://www.carecinstitute.org/wp-content/uploads/2022/11/EN_Post-Pandemic-Framework_Background-report.pdf.

International Institute for Sustainable Development (IISD). 2022. *Just Energy Transition Partnerships: An Opportunity to Leapfrog from Coal to Clean Energy*. https://www.iisd.org/articles/insight/just-energy-transition-partnerships.

National Bureau of Economic Research (NBER). 2022. *Climate Change Around the World*. https://www.nber.org/papers/w30338.

World Meteorological Organization (WMO). *Early Warnings for All Initiative*. https://wmo.int/site/wmo-and-early-warnings-all-initiative.